BREEDING RABBITS

A COLLECTION OF HELPFUL ARTICLES ON HINTS AND TIPS FOR RABBIT BREEDING

BY

VARIOUS AUTHORS

Copyright © 2013 Read Books Ltd.
This book is copyright and may not be
reproduced or copied in any way without
the express permission of the publisher in writing

British Library Cataloguing-in-Publication Data
A catalogue record for this book is available from the
British Library

Contents

CUNICULTURE (RABBIT FARMING) 5

HOW TO RAISE RABBITS FOR FOOD AND FUR. ... 9

KEEPING POULTRY AND RABBITS ON SCRAPS. ... 25

MEAT FROM YOUR GARDEN - A HANDY GUIDE TO TABLE RABBIT KEEPING. 37

MODERN RABBIT KEEPING. 52

RABBIT KEEPING. 68

SMALLHOLDINGS - TWO ACRES AND FREEDOM. 80

SOME INTENSIVE SIDELINES OF FARMING. 86

THE PRACTICAL RABBIT KEEPER. 91

Cuniculture (Rabbit Farming)

Cuniculture is the agricultural practice of breeding and raising domestic rabbits, usually for their meat, fur, or wool. Some people however, called *rabbit fanciers*, practice cuniculture predominantly for exhibition. This differs from the simpler practice of keeping a single or small group of rabbits as companions, without selective breeding, reproduction, or the care of young animals. The distribution of rabbit farming varies across the globe, and while it is on the decline in some nations, in others it is expanding.

Domestication of the European rabbit rose slowly from a combination of game-keeping and animal husbandry. Among the numerous foodstuffs imported by sea to Rome during her domination of the Mediterranean were shipments of rabbits from Spain, they then spread across the Roman Empire. Rabbits were kept in both walled areas as well as more extensively in game-preserves. In the British Isles, these preserves were known as warrens or garths, and rabbits were known as coneys, to differentiate them from the similar hares (a separate species). The term warren was also used as a name for the location where hares, partridges and pheasants were kept, under the watch of a game keeper called a warrener. In order to confine and protect the rabbits, a wall or thick hedge might be constructed around the warren, or a warren might be established on an island.

Rabbits were typically kept as part of the household livestock by peasants and villagers throughout Europe. Husbandry of the rabbits, including collecting weeds and grasses for fodder, typically fell to the children of the household or farmstead. These rabbits were largely 'common' or 'meat' rabbits and not of a particular breed, although regional strains and types did arise. Some of these strains remain as regional breeds, such as the *Gothland* of Sweden, while others, such as the *Land Kaninchen*, a spotted rabbit of Germany, have become extinct. Contrary to intuitive sense, it was the development of refrigerated shipping vessels that led to the eventual collapse of European trading in rabbit meat. Such vessels allowed the Australians to harvest and more importantly, sell their over-population of feral rabbits.

With the rise of scientific animal breeding in the late 1700s, led by Robert Bakewell (among others), distinct livestock breeds were developed for specific purposes. Rabbits were among the last of the domestic animals to have these principles applied to them, but the rabbit's rapid reproductive cycle allowed for marked progress towards a breeding goal in a short period of time. Additionally, rabbits could be kept on a small area, with a single person caring for over 300 breeding does on an acre of land. This led to a short-lived eighteenth century 'boom' in rabbit breeding, selling, and speculation, when a quality breeding animal could bring $75 to $200. (For comparison, the average daily wage was approximately $1.00.) The final leg of deliberate rabbit breeding – beyond meat, wool & fur - was the breeding of 'fancy' animals as pets and curiosity. The term 'fancy' was originally applied to long eared 'lop' rabbits, as the lop rabbits were the first rabbits bred for exhibition. They were first admitted to agricultural shows in England in the 1820s, and in 1840 a club was formed for the promotion and regulation of exhibitions for 'Fancy Rabbits'.

In 1918, a new group formed for the promotion of fur breeds, originally including only Beverans and Havana breeds – now known as the 'British Rabbit Council.' In more recent years and in some countries, cuniculture has come under pressure from animal rights activists on several fronts. The use of animals, including rabbits, in scientific experiments has been subject to increased scrutiny in developed countries. Meanwhile, various rescue groups under the House Rabbit Society umbrella have taken an increasingly strident stance against any breeding of rabbits (even as food in developing countries) on the grounds that it contributes to the number of mistreated, unwanted or abandoned animals. Some of these organizations have promoted investigation and prosecution of rabbit raisers on humanitarian concerns.

Another important factor has been the growth of homesteaders and small holders, leading to the rise of visibility of rabbit raisers in geographic areas where they have not been present previously. This has led to zoning conflicts over the regulation of butchering and waste management. Ironically, many homesteaders have turned to cuniculture due to concerns over commercial rearing of other animals, namely cows, chickens and pigs, as well as a desire for more self-sufficient living. Conflicts have also arisen with House Rabbit Society organizations as well as ethical vegetarians and vegans concerning the use of rabbits as meat and fur animals rather than as pets. The specific future direction of cuniculture is unclear, but does not appear to be in danger of disappearing in any particular part of the world. The variety of applications, as well as the versatile utility of the species, appears sufficient to keep rabbit raising a going concern across the planet.

HOW TO RAISE RABBITS FOR FOOD AND FUR.

Frank G Ashbrook

BREEDING

NEXT in importance to selection of stock is the judicious mating of the rabbits. It is relatively easy to produce rabbits, but to produce them of a uniform type that will build up and advance the breed requires persistent care and continuous effort. One ideal must always be uppermost in mind and the rabbits must constantly approach nearer to that one ideal if the producer is to register real progress. A uniform product depends upon such knowledge and control over heredity that matings may be made with the assurance that the offspring will be of the certain definite type demanded. The more thoroughly a producer studies breeding practices and the habits and conformation of his rabbits the more closely he may approach a desired degree of fixity in breeding operations. Meat and fur of excellent quality backed by a recognized breed always true to type, will go far to insure success.

AGE TO BREED

The proper age of bucks and does for the first mating depends on breed and individual development. The smaller breeds develop more rapidly and are sexually mature at a much younger age than the medium weight or giant breeds. Does should be mated when coming into maturity. Some difficulty may be experienced in getting them bred if mating is too long delayed. On the average, the smaller breeds may be bred when the bucks and does are 5 to 6 months old; the medium-weight breeds at 7 months; and the giant breeds at 9 to 12 months. Some individuals within a breed will develop more rapidly than others; therefore, in determining the proper time for the first mating, maturity of the individual is more important than age.

GERM CELLS AND FERTILIZATION

The female egg cells, which are microscopic in size, develop and are released into the Fallopian tubes and uterus through ruptures occurring in the walls of the ovaries. In the rabbit, several egg cells are usually released at one time; consequently, the size of the litter is determined by the number that mature and are fertilized at a given period, and develop to birth. Reproduction begins when the egg cells are fertilized by the male sperm cells. These newly formed bodies, or fertilized eggs, become attached to the walls of the uterus, where they develop.

At each mating, a vigorous normal buck deposits many thousands of sperm cells. The excessive number produced is a provision by nature to insure fertilization, for only one sperm cell unites with one egg cell. Consequently, more than one service to supply additional sperm cells is not necessary, and if some other condition prevents conception, two or more services will not overcome the difficulty. Moreover, there is a distinct disadvantage in allowing more than one service, for excessive use lowers the buck's vitality.

BREEDING SCHEDULE

The breeding schedule to be followed will be determined by the type of production. It would probably be best not to attempt to produce more than two or three litters a year in raising animals for show purposes. The time for matings should then be so arranged that the offspring will be of proper age and development for the show classification. In commercial production for meat and fur, the breeding animals should be worked, if possible, throughout the year. With the gestation period 31 or 32 days and the nursing period 8 weeks, this requires mating the does at the time the litters are weaned.

If no "passes" (failures to produce young) occur, it is thus possible to produce four litters in a 12-month period. If the size of the litter is materially reduced for any reason, the doe may be rebred earlier than called for by the regular schedule.

The condition of the individual animal should be used as the index for the proper time of mating. If, upon weaning the litter, the doe is reduced materially in physical condition, she should be allowed to rest until she has regained proper breeding condition (Figure 30).

MAKING MATINGS

Does give evidence of being ready for first mating by restlessless, nervousness, efforts to join other rabbits in nearby hutches, and rubbing the chin on the feed mangers and water crocks. This condition continues for some time, and as the rabbit has no regularly recurring heat period, matings may be made over a period of time, provided the does are in proper breeding condition and not diseased or in molt. Before mating, both the doe and the buck should be examined to make sure that they are free from disease.

FIG. 30. TWO HANDSOME SPECIMENS OF THE CHINCHILLA BREED

The doe should always be taken to the buck's hutch for service. Difficulty will often be experienced if this procedure is reversed, because the doe is very likely to object to another rabbit being placed in her hutch and may savagely attack and injure the buck. Bucks are slow also in performing service in a strange hutch. Mating should occur almost immediately on placing the doe in the buck's hutch. After the buck mounts and falls over on his side, mating is accomplished, and the doe should be returned to her own hutch.

Occasionally it may be difficult to get a doe to accept service. In such cases it will be necessary to restrain her for mating purposes. To restrain the doe, the right hand is used to hold the ears and a fold of the skin over the shoulders; the left hand is placed under the body and between the hind legs.

The thumb is placed on the right side of the vulva, the index finger on the left side, and the skin pushed gently backward. This procedure throws the tail up over the back. The weight of the body is supported by the left hand, and the rear quarters are elevated only to the normal height for service.

Bucks accustomed to being handled will not object to such assistance. It is well also to hold the doe in this way the first few times a young buck is used. This practice will expedite matings and insure ready service in difficult cases.

With a little patience and practice the breeder can so develop the technique under this system as to insure 100-percent matings. This does not mean, however, that all the does will "kindle," that is, give birth to young, but it will help materially in increasing the percentage of those that will, for a great many matings will be made that otherwise would not have been accomplished.

One buck should be maintained for each 10 breeding does. Mature, vigorous bucks may be used two or three times a week for short periods. A breeding record should be made showing date of mating and names or numbers of buck and doe.

GESTATION PERIOD

The gestation period, or the period from mating to kindling, is 31 or 32 days. A very small percentage of litters may be kindled as early as the twenty-ninth day or as late as the thirty-fifth, but 98 percent of the normal litters will be kindled between the thirtieth and thirty-third days.

FACTORS THAT LIMIT CONCEPTION

Among the causes of failure to conceive are false pregnancy and sterility; and some of the factors that result in a low percentage of conception are extreme ages, poor physical condition, mating in fall, sore hocks and injuries, disease, and molting.

AGE

Young does may not be sexually mature at the time of service, and old does may have passed their period of usefulness and fail to conceive. The first mating should not be attempted until the does are sexually mature and properly developed. The proper age for first mating has been indicated under the heading "Age to breed." Does should reproduce satisfactorily as long as they maintain good physical condition and satisfactorily nurse their litters. In commercial herds, does properly cared for should breed until they are 2 1/2 to 3 years old. Occasionally, individuals may reproduce satisfactorily for 4 to 6 years.

PHYSICAL CONDITION

Does and bucks that are either abnormally fat or thin will have their breeding powers impaired materially or may become temporarily sterile. The condition should be corrected by adjusting the ration and delaying breeding until the animals are in proper condition (Figure 31).

SEASON

Early spring is the normal breeding season for the rabbit; consequently, a higher percentage of conception will occur at this time of the year than at others. At the United States Rabbit Experiment Station, the highest percentage of conception occurs during February and March and the lowest in August, September, and October.

SORE HOCKS AND INJURIES

Sore hocks and injuries that affect a rabbit's vitality should be corrected before mating is attempted. When the does are out of condition, the percentage that conceive will be low.

FIG. 31. A UNIFORM FAMILY OF NEW ZEALAND WHITES

DISEASE

Rabbits should never be mated when they show any symptoms of disease. Remove such animals from the herd and hold them in quarantine until they have completely recovered.

MOLTING

Molting is normally in fall, and the percentage of conceptions occurring then is small. At this season rabbits are low in vitality, because of the heavy spring production, the heat of summer, and the additional strain of molting.

The feeding and management practices throughout the year will have an influence on breeding during the molting period. Adequate and properly fed rations will keep the rabbit in the best possible condition, and the molting period in well-fed animals will be much shorter than when the ration has been unsatisfactory. Proper feeding will assist the rabbits again to attain good physical condition, and when they are in full coat many breeding difficulties will be automatically overcome.

STERILITY

Occasionally a sterile rabbit will be found, and other individuals may be rendered temporarily sterile by one or more of the factors already discussed. The breeder should study each case carefully and if possible remove the cause of failure to conceive. Individuals that fail to respond to treatment should be discarded.

FALSE PREGNANCY

Does may be bred or stimulated sexually and shed the egg cells but fail to become pregnant. False pregnancy may be due to an infertile mating or to a sexual excitement caused when one doe rides another. Whether riding or ridden, does may become "false pregnant" and be unable to conceive until the false-pregnancy period is over. The period lasts for 17 days. After 18 to 22 days the doe may give evidence of the termination of false-pregnancy by pulling fur and attempting to make or build a nest.

TEST MATING

Test mating is the returning of the doe to the buck's hutch at stated intervals to determine whether she has conceived. If on being placed in the hutch the doe "growls" and avoids the buck, it is a fairly good sign that she is pregnant.

The breeder of show animals who is making matings for kindling at a definite time and the commercial breeder who is interested in keeping his does working as much of the year as possible can use the test mating system to good advantage.

All does should be test-mated when they pull fur and attempt to make nests 18 to 22 days after mating (false-pregnant does), when they prepare their nests several days in advance of the correct time for kindling and do not keep the nests clean, and when they fail to take on flesh or to show signs of pregnancy.

In view of the fact that a number of does that are bred and fail to conceive may experience false pregnancy, test mating on the eighteenth day after mating will be likely to detect the largest number of does that have failed to conceive. Does may also be test-mated at other times, and it may pay to test-mate a few days after mating as well as on the eighteenth day.

INBREEDING

In response to many inquiries as to whether inbreeding is desirable; that is, whether rabbits that are closely related should be mated, the average rabbit raiser is advised not to attempt inbreeding, for the following reasons:

Inbreeding knows no favorites. It will intensify poor qualities just as readily as it will good qualities.

The average breeder is unable to judge exceptional qualities in his breeding stock and usually does not have the necessary knowledge of the previous history of his animals to know what results may be expected.

Because the rabbits of the average breeder are usually of mixed inheritance, inbreeding such animals will always result in a variety of progeny.

Inbreeding is not harmful in itself, but it is sure, rapid, and effective in revealing the genetic structure of living forms. It will always remain a most potent procedure in developing and improving any breed of rabbits; in fact, no procedure other than close mating with rigid selection can be relied upon unfailingly to fix a type. Inbreeding, however, is a two-edged sword with which the ordinary rabbit raiser cannot afford to play. Discarding all undesirable forms, which is a necessary part of inbreeding, requires courage and considerable financial resources.

LINE BREEDING

Line breeding is the same in principle as inbreeding, except that the matings are made with animals that are not so closely related. Consequently, the characteristics of mated individuals, whether desirable or undesirable, are not fixed in the offspring so rapidly as when inbreeding is practiced. Probably most attempts by the novice at inbreeding or line breeding are made to avoid purchasing a new buck. Rather than take chances of obtaining inferior offspring by making close matings, it would be better for the novice to purchase a new buck of the desired type when it is necessary to breed does that are related to the herd buck.

CROSSBREEDING

Crossbreeding is the practice of mating a purebred rabbit of one breed with a purebred rabbit of another. This form of breeding is adapted principally to the production of new strains and should be attempted only by breeders with considerable experience.

IMPORTANCE OF HEREDITY

A few fundamental principles of breeding evolved from years of scientific study and observation should be noted carefully. Present evidence indicates that environment has little to do directly with improvement of animal form and that proper care and management practiced over several generations have no cumulative effect in developing a better breed of rabbits. Good feeding and care do, however, have the indirect value of providing a basis on which to select individual rabbits that because of their heredity will respond most satisfactorily to such care and management.

If improvement in rabbits is brought about, it must come chiefly through the hereditary factors transmitted through the germ cells. Effort, therefore, should be concentrated toward improvement by so mating the animals as to recombine these factors in more desirable forms.

The facts in brief are: Thousands of factors, called genes, determine the inheritance of each individual. The genes are collected in groups like beads on a string or like little packets. The groups are called chromosomes. These are of microscopic size, and the number in each cell is definite for each species. In the rabbit the number of chromosomes is 44, made up of 22 pairs. One of each pair comes from the father and the other from the mother.

The numerous breeds of rabbits, differing in size, color, and form, have resulted from various more or less stable combinations of these chromosomes. Frequent new combinations account for off-type individuals cropping out within a breed and explain also the extreme variability noticeable in the newer breeds as compared with older breeds that have eliminated most of the variable factors. The chromosomes frequently exchange genes, and this regrouping permits various recombinations, which are at once the hope and the despair of animal breeders. Without such variations there is no chance of improvement; with them there is no assurance of fixing a type without constantly selecting animals with desirable factors and discarding those with undesirable ones.

Another form of variation, known as mutation, though less common and less important from the breeder's standpoint, has produced types of some of the most important commercial varieties of rabbits. An example is the rex type, in which the guard hairs are either shorter than the underfur or entirely absent. Rex is recessive to the normal coat, and consequently a normal-haired rabbit may possibly be a carrier of rex. Any breed can be "rexed" within three generations by proper matings, if a sufficient number of rabbits are produced. When the offspring of a normal-haired rabbit and a rex-appearing animal are bred together, 25 percent of the litter will be pure rex, and if those of a New Zealand White and a Castorrex are mated, about 1 out of 16 of the offspring will be both white and rex.

Woolly, or long hair, is another mutation in rabbits. This, however, unlike the rex, is an undesirable trait. Woolly in rabbits is also a recessive, and consequently any rabbit that one suspects of having the woolly character can be tested by mating it with a woolly-appearing rabbit (and therefore pure for this character). If any young rabbits produced from this mating show the woolly character, one can be certain that the animal being tested is a carrier of woolly even though it appears normal-haired. Neither should be used in breeding. This particular type of woolly is different in appearance from that of the Angora rabbit.

KEEPING POULTRY AND RABBITS ON SCRAPS.

Claude H Goodchild

BREEDING AGE

The satisfactory age at which a doe can be bred from, depends on how it is matured. Most varieties will conceive at 4 to 5 months. It is dangerous to breed from immatured stock. They will not mature later, prove bad mothers, and produce valueless progeny. In summer months maturity can be at 6 months, in the winter 7 months – much depends on growth.

Best results are obtained by starting with stock at least 8 *months of age.* Does are liable to neglect their first litters, especially in the autumn and winter.

Usually this defect does not re-occur. This may be due to moult, lack of sunshine, and suitable foods to stimulate a flush of milk. Experienced mothers are the best to breed from during the more difficult period, mated to a young, vigorous buck.

QUALITY OF BREEDERS

It is essential to start with the best quality healthy stock. On no account breed from a deformed or unhealthy rabbit. Chief indications of a healthy rabbit are, shiny smooth coat, alert eye, clean nose, dung in firm pellets.

HANDLING

A rabbit should be lifted by supporting the hind quarters with the left hand and gripping the ears and fur at nape of neck with the right (*see Plate* 44). Avoid handling doe in kindle, especially just before the due date.

STUD BUCK

The buck or bucks used for mating the does should be of the best quality. Extra care and good regular feeding is necessary. The buck imparts health and strength to the litter, especially in the early stages. Never breed from a buck which is out of sorts. A stud buck should not be allowed more than 5 matings per week.

MATING

There is no definite period when a rabbit is in season. It varies with the time of the year and condition of the rabbit. In the spring, does are practically always ready to mate, whereas at other times of the year it may take several days to gain success. Matings in winter are best obtained in warmer days. Place the doe in the buck's hutch, when one of four things will happen.

1. The doe sits hunched up and ignores buck.

2. The doe chases round the hutch in playful manner.

3. The doe whines or goes for the buck.

4. The doe remains still, arches her back and makes things easier for the buck.

In cases 1 and 3 remove doe and try again later. 2. Wait a few minutes and case 4 may occur. 4. Successful mating is almost a certainty.

The completion of the act is shown by the buck falling over on its side with ears back with or without crying out.

A scream is thought by beginners to indicate hurt to the doe. This is not so but perfect mating.

Sometimes a buck will do this part without effecting an entry.

If this is suspected, examine that part of the doe for signs of the conception. It is not always necessary for visible satisfactory matings to be successful, and it is advisable to place the doe in with the buck again on the 14th day.

An expert can tell at this period by hand handling if the doe is in kindle. A beginner should not try this. Damage can be caused by rough handling.

GESTATION PERIOD

A litter is produced 31 days after mating.

MATED DOES

Does can be purchased ready mated, but it is advisable to have a buck or one within easy reach for future mating. Avoid sending for mating by rail. There is so much loss of time and expense, and the journey tends to increase the risk of unsuccessful mating.

KINDLING

Does are usually very restless several days before the litter appears. To avoid disturbance during this period, thoroughly clean the hutch out one week before kindling date, and give a liberal supply of litter in the hutch. Usually this is all heaped in a corner quite a time before the litter arrives.

Does often go off their feed just prior to kindling. Remove any stale food and give fresh. Fresh water is essential.

ARRIVAL OF LITTER

Although does have usually made a large heap of litter in the darkest corner of the hutch for the expected family, they usually produce the litter before completing the final act of pulling quantities of fur from their flanks and back to keep the young absolutely warm.

Some careless mothers produce their litter scattered all over the hutch floor, and in the cold weather they will be dead in a very short time. This neglect is unaccountable and usually does not recur.

If this is noticed in time, put all young together in a nest surrounded with warm fine hay. The mother will later do the rest. (*See section referring to Mothers eating young.*)

SIZE OF LITTER

The number of young in a litter varies considerably. Anything from 1 to 14. An average litter is 5 to 7.

It is always advisable to examine the young as soon as born – removing any dead in nest, if left they may all die. If litter is too large kill off surplus, allowing the strongest babies to live.

The number to leave depends on the capabilities of the mother which are regulated by the time of the year and quality of food. From January up to the end of July, 6 to 7 is a good litter to leave, in the other period 5 to 6.

FOSTERING

If you have a doe of special merit and prize the progeny, mate another doe the same day as her and assuming the litter of your special doe is of average size or over, kill off the young of the other doe, and put half the young of the special doe to this doe which is now termed a foster-mother doe. Before doing this operation, take note of the following instructions:

1. Take both does concerned out and feed.

2. Make sure your hands smell correctly by rubbing fingers in litter of hutch concerned before handling young.

3. Take away all foster-mother's young—don't mix.

4. The young which are put into the foster's nest can be made to smell correctly by sprinkling with litter from foster's hutch.

5. Feed does with 'tit bits' as soon as put back to distract attention.

6. Disturb nests as little as possible, and make sure there is no fur on the young when transferring.

Fostering can be done up to 4 days, but the safest time is between 12 and 24 hours after birth. A difference of age between the young of the doe and foster of more than 2 days makes it tricky. The quality of the milk changes after the birth of the young.

WEANING

The age at which a rabbit can be weaned depends on the time of the year and growth. When food is plentiful and good, weaning can be done at 4 to 5 weeks, in autumn and winter

6 to 7 weeks. Great care in careful feeding should be exercised when young are first weaned. Give only the best food (as described later) and feed often.

If allowed to get hungry they are liable to over-gorge at their next meal and develop indigestion and scours.

It is advisable when possible to wean young into runs or colonies with several in a place where there is room for exercise.

RECORDS

It is advisable to keep full records of all stock intended for breeding. A system should be adopted to keep a trace of the breeding stock. If kept in single hutches a hutch card is all that is necessary.

The difficulty arises when mixing litters together, when some form of individual marking is necessary to keep trace.

TATTOOING

Tattooing is simple and most effective, and if done with care is lasting. Place the rabbit's ear flat on a piece of soft wood and with a large-sized needle dipped in marking ink, prick out the number or marking required.

After the operation, smear over with the ink.

Another method is to ring the stock. Special rings are supplied for this purpose. These are put on by slipping over the hind foot up to the hock. These are made in different sizes for each variety by the British Rabbit Council, 273, Farnborough Road, Farnborough, Hants.

BREEDING PERIOD

Tame rabbits breed freely throughout the year, but the output from a doe should be limited to 4 litters per year. It is not advantageous to go beyond that, or it will effect the health of the doe and render it unable to rear litters until it has a complete rest.

TABLE SHOWING STEADY BREEDING WITHOUT UNDUE STRAIN ON THE DOE

Mate	Born	Wean	Kill Youngsters
January	February	April	June to July
April	May	July	September to October
July	August	October	December to January
October	November	January	March to April

The output can be compared to a motor – good petrol and

oil, and steady driving gives lasting wear. Given good food and steady breeding the doe will last for several seasons. Abuse of this rule means complete breakdown.

A doe will remain good for breeding for an average of 4 seasons, but is usually replaced before that time by one of better quality.

INBREEDING

The extent to which the breeding of relations together can be conducted depends entirely on the stamina and type of rabbit.

For table purposes avoid inbreeding; it serves no good purpose. For exhibition stock inbreeding is necessary, but should be done very carefully.

Often certain outstanding qualities are only contained in certain closely related rabbits, and to improve the standard of the strain, the mating of close relations is necessary. Having done this once, the progeny should be mated to almost unrelated stock.

Disastrous results can be obtained by inbreeding at random. To be successful it is necessary to keep a strict record of the breeders and make full observations at all stages and avoid breeding from any stock which is lacking in stamina and vitality.

HAND REARING

It happens that a doe may die possibly from the effects of kindling or other defects, and usually breeders, chiefly out of sympathy, consider they can rear the young with a fountain pen filler and cow's milk.

It is not worth the trouble, even if they are valuable youngsters. Chances of success are too remote. If attempted, dilute cow's milk with 25 per cent to 50 per cent water, and feed morning and night. Avoid over-feeding. Snag – change of food causes indigestion, diarrhoea and death.

SEXING

The earliest stage when youngsters can be safely sexed is at 4 weeks. Great care should be taken in this operation. Avoid undue forcing in order to expose the organ.

Hold the rabbit by placing your thumb and index finger of your right hand at base of ears and the palm of your hand at back of the head, allow the rump of the rabbit to rest limply between your legs, then press the organ with thumb and first finger of the left hand. The male organ protrudes in a cylinder shape and the female only protrudes one side in a triangular shape. (*See Plate* 45.)

CASTRATION

Castration is very seldom done on rabbits. It has very few advantages. The chief reason is to prevent bucks from fighting in colonies. For table purposes only, bucks are killed before they reach that stage. For fur it has the effect of making the rabbit lazy and retards the fur growth. Castrated rabbits never appear in prime fur. For these reasons there is no object in describing this skilled operation.

MEAT FROM YOUR GARDEN - A HANDY GUIDE TO TABLE RABBIT KEEPING.

Walter Brett

MATING YOUR BREEDERS

Because the success or otherwise of your enterprise depends upon the production of a steady flow of healthy, robust youngsters, the mating of your breeding stock must not be a haphazard business. It must be properly timed and you must assure yourself that it is properly carried out.

Before we go on to tell you how to mate, however, there is a point to be settled: Are you sure of your rabbits' sex?

When you bought your rabbits you ordered so many does and so many bucks, of course, and probably the dealer told you at the time which were the does and which the bucks, but maybe you got them mixed when installing them in their hutches (rabbits are often very much alike) and are not now quite sure of them.

Attempts to "mate" two does or two bucks invariably lead to fierce fights, and sometimes serious injury, so you see that it won't do to take chances.

How to tell a Rabbit's Sex. Fortunately, it is easy to sex a rabbit.

Firstly, buck rabbits and doe rabbits are of appreciably different build. The buck has a much broader and shorter head than the doe. When you are familiar with rabbits you can tell the sex at a glance from the heads alone.

As a beginner, the head would not be a sufficiently reliable guide to you, however. You therefore obtain your information in the following manner:

Hold the rabbit on your knee in a sitting-up position, with the under part of the body uppermost. Grasp the ears firmly with your right hand to hold the animal still. Now, with the forefinger and thumb of your left hand, open the sexual organ.

In the female the vulva is oval-shaped. In the male the orifice is round. Other differences will also be discernible.

The Right Breeding Age. You should have bought rabbits eight months old. If you bought younger stock, you must wait a while before breeding from them.

Although rabbits are capable of breeding at three months old, the mating of such immature stock leads to the production of weakly litters and overtaxes the powers of the parents so that their development is retarded. Both does and bucks come to maturity for breeding at between five and eight months, and should not be mated before the latter age.

The Rabbit's Natural Breeding Season. All rabbits breed most freely in spring and summer, especially from March to June inclusive, when the weather conditions are favourable and there are abundant supplies of succulent greenfood.

However, if you keep your rabbits comfortable and reasonably warm they will breed at all seasons of the year, although not so freely in autumn (September to November inclusive) when most of them are moulting into their winter fur.

Five Years of Useful Life. A good breeding doe will have at least three litters a year—quite likely four—until about her fourth year, when she does not usually breed so readily. You will do well to replace five-year-old does with younger animals.

The average number of young rabbits per litter is six. Spring and early summer are more favourable for large litters and good growth of the youngsters than other seasons of the year.

Why you must keep the Sexes Separate. You have already been told you must keep your does and bucks in separate hutches, except, of course, when you decide to mate a particular pair.

There are many reasons for this. A doe in kindle as the result of running with a buck for some time is very likely to destroy her young when they are born and to be lacking in good maternal qualities in future, while one which has refused service under these conditions is almost certain to become a "shy breeder" and to refuse the attentions of a buck to which she is introduced in the correct way.

All bucks over three months old should be kept in individual hutches. They rarely do well in batches on account of fighting. But young does may be kept together until ready for breeding if this is convenient.

To tell when a Doe is "In Season." A doe can only be mated when she is "in season." You can ascertain when "season" occurs in various ways. The most common sign is stamping with the hind feet, rolling, scratching in the corners of the hutch or carrying mouthfuls of hay or bedding.

If a doe is not in season the fact will generally be apparent when she is placed in the buck's hutch. She will either crouch in a corner of the hutch or "square up" to him and show fight. In either case you should immediately remove her to her own hutch and try her again next day.

Should you find a doe backward in coming into "season" try putting her in a *vacant* hutch previously occupied by a buck. If this doesn't have the desired effect, give her some maple peas which have been soaked for forty-eight hours in warm water until they have begun to sprout. Give a handful on each of two or three mornings.

It is a fact that maiden does are usually more difficult to mate in winter.

How to mate. For mating, place the doe in the buck's hutch, never the buck in the doe's. You should see that both partners to the mating are in good condition, especially the doe.

Introduce the doe quietly into the buck's hutch. If she is ready for mating she will willingly accept service almost immediately by stretching herself out and raising her hindquarters.

The buck should be allowed to serve the doe once only. When the service has occurred—and it is a matter of a few moments only—the buck will fall over on his side. No service is complete without the fall. When it has taken place, remove the doe and replace her in her own hutch.

WHEN THE DOES ARE "CARRYING"

When to expect the Young Rabbits. Most does give birth to their young on the thirty-first day after mating.

Occasionally, and with maiden does in particular, the litter may be born one or two days earlier. Alternatively, birth may be delayed for a day or two over the normal period, this being most common in older does.

Feeding the Carrying Doe. When the doe is in kindle (carrying her young) you should take especial care to give her as succulent a diet as possible. She should have the first choice of the greenfood if this is scarce. If you can manage a drink of milk she will greatly appreciate it. Milk powder is useful in preparing this drink when milk is not easy to obtain.

Novices sometimes get anxious because they notice that a doe getting near the end of her time takes to lying stretched out on her side instead of sitting in the usual way when at rest. This, however, is quite the usual posture towards the end of pregnancy.

A temporary falling off of appetite during the last day or two is also fairly common. You may take this as a pretty good indication that the litter will be born within the next twenty-four hours.

Provide a Nest-box. Three weeks after mating you should provide the doe with a nest-box. This must be big enough for the doe to lie in. A good size is 16 in. long by 12 in. wide, and 6 or 8 in. deep.

One of the principal purposes of the nest-box is to protect the young so that they may not stray from the nest or be dragged out, thereby running the risk of dying from cold.

Just before the litter arrives you will see the doe to be busy with the nest-box. She will be carrying mouthfuls of nesting material to it, and will be lining it with fur stripped from her chest, flanks and belly. This stripping of the fur serves another purpose; for if has the effect of baring her teats, and thus allowing the baby rabbits to have easy access to them.

Necessary Care to prevent Mishaps. In addition to careful feeding, you must exercise all necessary precautions to ensure that the doe is not frightened whilst she is carrying. Keep dogs and cats away, children also, and walk quietly when in the vicinity of the hutch.

Two or three days before the litter is expected, clean out the doe's hutch thoroughly, remembering that this will be the last opportunity you will have to do so for some little time. You may wish to remqve the doe from her hutch when cleaning. Do so without causing fright or upset. Grasp her ears with one hand and place the other under her haunches so as to support her full weight.

WHEN THE LITTER ARRIVES

On the day the litter is expected, keep right away from the hutch, merely providing a little fresh food in case it may be needed and also making perfectly sure that the supply of drinking water is ample. Rabbits need water more at kindling time than at any other. If their thirst is not quenched, then the fever and consequent discomfort increase and lead to such nervous excitement that a doe may turn on her young, as you have learned, and mutilate or even devour them.

Don't Peep. Don't be inquisitive when you suspect the litter has been born. Don't take a look-see. The doe will take about an hour in which to give birth to her litter and settle down. If all is well afterwards, she will probably be found sitting quietly in her hutch some distance away from her nest, which will appear as a mass of fur. If you watch carefully, you will see the nest heave gently; that is a sign that the young are alive. Rest content with that sign and don't attempt to confirm it by actual inspection of the nest.

The only occasion when you need interfere is when the doe has been scared or disturbed while kindling and you find things rather upside down in the hutch, with the youngsters scattered about the floor instead of being carefully covered over in the nest. In this case collect the little naked things tenderly and put them back in the nest, covering them over with the nesting material and a little of the fur. Smear your hands with the urine-soaked sawdust before touching the babies, this so as not to taint them with the human scent.

It is a toss-up even so whether the doe will settle down with the "made-up" nest, but it is your only chance of saving the litter. Young rabbits are born bare of fur and even a short experience of draughts and cold air may be fatal at this early stage.

How Soon you may look to see what the Litter is like. Some experienced breeders examine nests of young rabbits within a day or two of birth, but you had better let well alone and refrain from ascertaining how many there are in the litter until the young rabbits begin to come out of the nest. This will be when they are between two and three weeks old.

If there are any youngsters born dead, rest assured the mother will usually deposit them in the outer portion of the hutch, when you should remove them.

It may, however, be rather too much of a strain on the patience of the novice to wait for the fortnight or more to see what luck he has had. If your curiosity *is* too strong to be so long suppressed and you feel that you really *must* see the litter, you may examine the nest after 24 hours have elapsed. Take only the quickest peep then.

Of course, there may be some *real* reason to think that all is not right with the litter. If, for instance, no movement is seen in the nest for the first day or two you may reasonably suppose that something *has* gone wrong. In such cases it will be best to take a chance and see what has happened. Proceed as follows:

Remove the doe from the hutch and put her in a spare hutch or a box where she cannot see what you are doing. Then take out the nest-box, and carefully but quickly examine the nest and overhaul its occupants.

If all is well there is no need to do more than take a hasty look at the youngsters, notice how many there are and remove any dead ones before replacing the nest-covering and the nest-box and finally the doe.

If *all* the youngsters are dead the doe should be mated again without delay. A doe will take the buck readily if put with him within 48 hours of kindling.

Don't forget, before starting your examination, to rub your hands with a little of the damp litter from the soiled corner of the hutch. As already mentioned, this prevents the doe from noticing any strange scent about her nest or family.

A Doe Seldom neglects her Family. You may imagine that

because a doe is never seen to go to her nest she must be neglecting her family. You need have no anxiety on this score. A doe feeds her young only twice during the 24 hours, and nearly always chooses a time when nobody is about.

If you do happen to surprise a doe in the act of feeding her youngsters, go away at once, very quietly. The babies cling very tightly to their mother's teats and if she suddenly moves away from the nest, as she is very likely to do if disturbed, they will probably be dragged out.

Feeding the Suckling Doe. So long as the young rabbits remain in the nest they require no attention beyond that given them by their mother, who will feed them and keep them clean.

It is up to you, however, to feed the doe so as to ensure that her milk is sufficiently abundant in quantity and rich in quality to maintain a steady rate of growth for the youngsters from birth upwards.

When the doe's appetite increases, as it will normally about the second day after she has kindled, this should be met with a corresponding increase in the amount of food.

The best means of increasing the food of any healthy rabbit is always by an augmented ration of greenstuff. This applies especially to the nursing doe for two reasons: Firstly, because greenfood makes milk; secondly, because if a doe is largely fed on greenfood, her young, when they begin to feed for themselves, will come to no harm through eating quantities of their mother's food.

If a doe is allowed all the greenfood she can eat night and morning, and, in addition, a lump of mash—bread, boiled potatoes or peelings, house scraps, etc., dried off with broad bran if this is available—about the size of a duck's egg, she and her litter should do well.

EARLY LITTER MANAGEMENT

When the Litter is born in Winter. When litters are born in the winter, or when weather conditions are really cold, you must keep your doe well supplied with nesting material all the time the youngsters are in the nest, especially during the first few days, before they get their coats. They are born blind and normally with only a coat of fine down. By about the end of a week they will be comparatively well furred, therefore not so susceptible to chills, but nothing should be left to chance.

At about ten days old the babies will open their eyes.

This guarding against chills in a litter is really a necessary measure. A good brood doe can be trusted to protect her nest from cold if she has material available; and it is a common sight to find bedding and hutch litter piled over the nest almost to the height of the hutch roof in severe weather. You should, however, help by covering the front of the hutch with a thick sack on cold nights, and during the day, too, if the weather is wet, windy or very cold.

When Litters are born in Summer. In summer it sometimes happens that a young doe will show a desire to mate when her litter is still very young. As a consequence she may cause annoyance by neglecting and harassing her babies. When this is apparent you had better remove the doe and mate her forthwith, after which, if she is kept quiet, she will usually behave herself when replaced in her original hutch with her family.

Cleaning Out the Breeding-hutch. With a doe of normal habits it should not be necessary to clean out the hutch during the first three weeks after kindling, as one corner of the hutch only is used for sanitary purposes as a rule.

In this case cleaning may be confined to the occasional removal of damp hutch litter from the used corner with a shovel, the soiled material being replaced with fresh.

If, as occasionally happens, a doe is dirty in her habits, a more thorough cleaning will be necessary, as ammonia fumes from the urine will not only make the atmosphere of the nest unhealthy but may also give rise to serious eye trouble in the young rabbits.

Any cleaning out at this stage you had best do just before feeding time, when the doe is hungry. Remove her from the hutch and place her apart in a spare hutch or box with a little food to keep her quiet. Then, as quickly as possible, clean out the hutch, without if possible disturbing or interfering with the nest, and put in fresh litter and bedding. When all is snug and clean return the doe, giving her the usual feed at once.

When the Young will start to feed. Almost the first action of the youngsters, when they begin to come out of the nest, which will be when they are from fourteen to twenty-one days old, is to investigate the food provided for their mother. This does not mean that they are ready and able to shift for themselves, only that they are ready for solid food in addition to their mother's milk. Consequently, whilst you may allow them to eat anything the doe has been having, you should not introduce new foods at this stage.

At first the youngsters will merely nibble at the food, taking very little at a time. In a day or two they will "find their appetites" and by the time they are a month old they will be taking their full share.

Mating the Doe Again. A question which is important is how soon after kindling should a doe be mated again. Although a certain amount of difference of opinion exists on the matter,

Breeding Rabbits

it is safe to say that (except in the instance referred to earlier) you should not put a doe again to the buck until at least two weeks have elapsed after kindling. She must have time to build up her strength.

You should not mate a doe while she is moulting.

MODERN RABBIT KEEPING.

King Wilson

Breeding

Reproduction

Does should be mated when coming into maturity, for difficulty may be experienced in getting them to breed if mating is delayed too long. The proper age of bucks and does for the first mating varies with the breed. On the average, small breeds may be mated at 5 months, medium breeds at 6 months, and the giants at about 7-9 months.

Rabbits, unlike most other mammals, do not have a definite "heat" period, but will breed at any time except in the autumn. However, spring and summer are the natural breeding seasons, for in the autumn the rabbits moult into their winter coats and do not breed freely. Bright, warm, sunny weather is most favourable; an improving standard of nutrition and body condition increases the chances of conception and improves the size of the litter. Table-rabbit producers, who want stocks to breed in winter, should provide fresh cabbage, kale or other green food in addition to roots. At this time it is also advisable to feed liberal amounts of concentrated foods, such as cereals, which are rich in vitamin E (fertility vitamin). During the spring and summer, does often pluck their fur.

The doe should always be taken to the buck's hutch for service for, if the reverse procedure is adopted, the doe may turn on the buck. Bucks are also slow in serving in a strange hutch. Mating usually takes place within a few minutes. A mating is indicated by the buck falling backward or sideways off the doe and grunting. Should there be any doubt about the service, a second union may be permitted before the doe is returned to her own hutch. If no mating occurs, the doe should be returned to her hutch and another attempt should be made to mate her about three days later. It is unusual for a pair not to mate on the third day unless one or both are too old or too young. An adult doe that has not bred for a year or more may be a reluctant breeder and, before being mated, she should be placed for a day in a hutch previously occupied by a buck. A few does do not readily mate because they are too fat; this condition can be remedied by giving them exercise and more bulky food.

One buck should be kept for about ten breeding does. Mature vigorous bucks may be used two or three times a week.

Breeding Rabbits

The small domestic breeder can usually arrange to have does served by an outside buck on payment of a small stud fee. A breeding record should be kept showing date and number of animals mated.

After mating, the doe should be kept in a quiet place in her own hutch and given plenty of food, especially greenstuffs and mash. A plentiful supply of good hay or straw should be provided for nest-making and care should be taken to ensure that the hutch is dry and free from drainage from the pens above; top-tier hutches are safest if the hutches are not quite watertight.

Pregnancy normally ranges from 30 to 33 days, the average being 31 to 32 days. False pregnancy sometimes occurs, more particularly in does that have not been mated for a considerable period. All does should be test-mated when they pull fur and try to make nests some 18 days after mating. Test-mating on the 17th-18th day after mating helps very materially to detect the majority of does that have failed to conceive at the first mating. Second matings on the 18th day with non-pregnant does generally prove successful.

The breeding does at this stage should be kept well away from bucks, which might disturb them by stamping and cause the does to scatter their young.

The number of young in the litter varies with the breed

and, as a general rule, large breeds have bigger litters than small breeds. In the utility breeds the average litter size is about five to seven but larger litters are quite common. The first litter is usually smaller than the second and the numbers tend to diminish after the eighth litter. Three litters a year is a satisfactory average and these should be obtained in the first eight months of the year as the youngsters born in the first part of the year grow better than those born later. The breeding results in September and October are usually poor. Does are at their best as breeders during the first three years and it is not usually economic to keep them for more than three breeding seasons.

Angora does should be clipped before mating as long wool in the nest may get tangled round the young and strangle them.

Fostering

The beginner is advised not to interfere with the young. As experience is gained it is a good plan to mate several does at the same time so that some of the young from the large litters can be fostered on to does with small litters. For commercial purposes 6 to 7 is the best number to leave with a doe. Fostering is best done by the day after birth. Before touching the nest, the hands should be washed to remove any odour and then rubbed in some familiar foodstuff or in the bedding. During the process of fostering, the attention of the doe should be distracted from the operation by feeding attractive green food. The centre or top of the nest should then be carefully opened, without breaking the side, and the young examined. After examination, the next step should be the transfer of some young from the large to the small litters; the nests should be carefully re-covered and left as they appeared at the outset. An aggressive doe may have to be removed from her hutch but this precaution is not usually necessary if green food is placed well away from the nest. The nest should never be touched by unwashed hands that have been in contact with dogs, cats or other animals, including other rabbits, as does have a very keen sense of smell.

Sexing at Birth

There is no known method of controlling sex. On the average slightly more males than females are born. Sexing of the young is generally done at weaning or at 3 months but it is possible for those with good eyesight to determine the sex at birth or at 3 days (Plate IV), by the aid of a watch-maker's glass.

For sexing at birth the baby rabbit should be held gently in the left hand, with its head towards the operator and with the hind legs extended. Pressure should then be gently exerted on each side of the vent, using the thumb of the right hand. In the male the penis appears as a rounded tip whilst the female organ is slit-like and without a rounded tip. The distance between the anus and vent is somewhat greater in the male (Plate V), and there are a pair of reddish brown specks near the vent which are absent in the female.

At weaning time the chief distinction is that when the vent is gently pressed open the aperture in the case of the male remains circular whilst it becomes slit-like or V-shaped in the female.

Growth

The growing embryo remains relatively small until the end of the third week but after then the rate of growth is rapid and it is desirable to give the doe better food during the last 10 days of pregnancy. When she has kindled she should be given a feed of bran and thereafter the food supply should be increased according to appetite. After the first two weeks, during which the young are entirely dependent on the mother's milk, they begin to run about and to nibble at the mother's ration. The young may remain in the nest up to the age of 3 weeks if they receive a good supply of milk. At 2 to 4 weeks, the doe produces 3 to 5 oz. of milk daily if properly fed. Rabbits' milk is exceptionally rich in growth-promoting substances; in comparison with cows milk it is about four times as rich in protein and fat and three times as rich in minerals.

The young double their birth weight in the first week and after they begin to eat solid food the growth rate is very rapid, especially from 3 to 8 weeks. As a rough guide, the increase is about 1 lb. per month with a tendency for greater gains during the first 3 months and smaller gains later. The rate of growth varies widely with the breed, number of young, time of year, age of mother and food supply. Small litters of three or less grow more rapidly and are preferred for show purposes but a litter size of 5 to 8 probably gives the best results for utility purposes.

With litters of six or more, and especially with large breeds, it is desirable to feed rations that are somewhat above the average in protein and minerals. If the diet is deficient, the mother draws on reserves from her own body and this tends to impoverish the next litter. Symptoms of deficiency behaviour include urine drinking which indicates thirst, wire-licking which suggests shortage of water or minerals, and licking the limewashed walls which indicates calcium deficiency.

Exercise is essential for growing stock and large litters which are often cramped for space should be given a daily run on the floor. Large hutches, though more expensive than small ones, are not an extravagance as they yield good returns in the form of improved growth. Rabbits of 3 to 3 1/2 months generally get separate hutches.

The colony system of rearing up to 20 to 30 young in an empty shed generally ensures plenty of exercise. The young are not weaned until 6 to 7 weeks, and they are then sexed and put in the colony houses. By 4 months the bucks are usually housed in separate cages, as they require rather careful watching in colonies; bullies should immediately be removed to separate hutches. Undersized rabbits will do better if taken out and housed in hutches but, for the bulk of table-stock, colony rearing is usually easier than hutch rearing. It must be remembered, however, that if disease breaks out, it spreads more rapidly in colonies.

The three commonest faults made in rearing are, firstly, a food pot that is not large enough to enable the whole of a large-sized litter to eat together; the second is insufficient food, and the third is overcrowding, especially with large litters. These points should receive careful attention for whatever the reason, stunted rabbits are difficult and costly to bring up to pelting condition.

There is considerable risk of scouring, debility, unthriftiness and high mortality amongst youngsters from one to three months old if does and litters are kept on poor quality foods. Litters appear to develop fairly satisfactorily for some weeks when kept in confinement on a diet of rough grass, cabbage and bran but between five and ten weeks they collapse and die on account of deficiencies in the diet. Similar troubles occur in youngsters grazed too frequently in Morants on small lawns but part of this trouble may be due to the ingestion of parasites which may come from dogs infected with tape-worms.

The most serious disease is coccidiosis against which, however, the following treatment has proved effective: Sulphamezathine is added to the dry food at the rate of 1 per cent but, if it is to be used satisfactorily in controlling outbreaks, it must be given in the early stages, e.g., within about 2 days of infection, usually when the rabbit is 1 month old, for about 3 to 5 days. Repeat if necessary by weaning. In very mild cases slightly smaller doses are helpful but do not prove quite so satisfactory. The treatment can also be used for adults although the effects are not so apparent. Large-scale rabbit keepers (and poultry-keepers) can obtain sulphamethazine on request, but the domestic rabbit keeper will need a prescription for the drug from a veterinary surgeon.

In various experiments, up to 80 per cent of the youngsters fed on a diet of green food and hay but no concentrates died. The addition of 3/4-1 lb. of cooked potatoes per day to medium-sized does with litters brought down the mortality to normal and gave as good results as a daily supplement of 4 oz. oats. Smaller quantities of cooked potatoes did not, however, prove as effective in lowering mortality as concentrates.

It is advisable to feed mixtures of green food instead of excessive amounts of any one kind. Green food provides valuable amounts of vitamin A, so essential for the normal growth and health of young rabbits. Care must, however, be exercised in feeding green food in wet weather.

Food is utilized most efficiently during the period of maximum growth, i.e., from birth to 4 months, when it is essential to give foods which are rich in all the essential materials for growth. After 4 months, as the rate of growth slows down, good results can be obtained with less concentrated foods than those fed during the earlier stages of life.

The future breeding stock should be selected from litters which show good power of survival and a rapid rate of growth. A quick growing, hardy strain is most desirable and it is sound policy to discard the stock from those strains whose chances of survival are poor and which show slow or uneven growth. Does and bucks which do not give the right type of progeny should be discarded as soon as they come into full coat.

Fur or Wool Growth

The rabbit has usually three moults in its first year and an annual moult in the autumn of each subsequent year. The baby coat is shed at 4 to 8 weeks and the second coat around 2 1/2 to 4 months. During the warm weather the coat of breeders is more or less loose, but the main moult occurs at the end of summer and early autumn. The fur breeder aims to produce good pelts from November to April, the best quality ones being obtained from December to February.

Breeders should choose their breeding stock from amongst those which moult cleanly and quickly, rejecting those rabbits which fail to come into full coat between two successive moults.

The feeding of stock on well-balanced diets helps good coat development. The addition of foods rich in oil, and more especially in linseed, improves the sheen or lustre of the coat. Milk is also excellent for promoting good fur development. Experiments showed that cod liver oil at a one per cent level was beneficial and in one experiment it increased the weight of coat by 17 per cent, although it proved injurious when continued over long periods. The addition of mineral mixtures to the diet of rabbits gave no benefit and in some cases it actually decreased the coat. The best way to ensure the necessary minerals for good coat development is to feed foods such as red clover and greens; as they are good natural sources of well-balanced minerals. Rabbits fed on a mixed diet containing bran did not produce such heavy coats nor such full, soft fur as those receiving a similar diet supplemented with a good mash. The substitution of bran for mash reduced the food costs but gave lower profits because of the poorer fur.

Cold weather encourages good coat growth but coat development is primarily a matter of heredity and a full-coated strain should be used for fur or wool production. The best coloured pelts are produced in shady hutches, for light and sunshine fade the colour.

Inheritance

During recent years great progress has been made in the scientific study of hereditary problems in the rabbit, and breeders are now able to make practical use of many facts which have been brought to light.

All the evidence indicates that environment plays its part in improving stock and that apparent improvements produced by better feeding and management have no cumulative effect in improving a breed. Good feeding and management are nevertheless important, for without them it would be difficult to select those strains which possess the most desirable characters. It is now generally accepted that hereditary factors are transmitted through the germ cell and that thousands of factors or genes determine the inheritance of an individual. The genes are collected together in groups inside minute microscopic bodies known as chromosomes. Each animal has its own particular number of chromosomes and the rabbit has 44, made up of 22 pairs. The ova of the female and the sperm of the male each contain 22 chromosomes, for during the process of development one of each of the pairs of chromosomes is lost. Fertilization restores the number to the original 44. Both parents share equally in determining the genetical make-up of the offspring.

Fur Colour

Many of the ordinary coat colours in rabbits are inherited according to the simple Dominant-Recessive rule of Mendelian inheritance. For example a true breeding black (Dominant) rabbit mated to a blue (Recessive) gives all black offspring. Outwardly these offspring (F1) are just like their black parent but when mated together they give, on an average, 3 black offspring to 1 blue. These blues breed true. Of the second generation of (F2) blacks, one on the average will breed true and the others will throw some blues like the F1 generation.

Impure blacks mated to blues give 50 per cent blues and 50 per cent blacks. With this type of test-mating a doubtful dominant mated to a simple recessive will give progeny of the dominant type only, if the dominant is pure. If the dominant is impure the offspring will only be about 50 per cent of the dominant colour.

The genetical make-up of rabbits for simple dominant and recessive characters discussed may be tested as follows:

A true-breeding black mated to a true-breeding black always produces all true-breeding blacks.

A true-breeding black mated to an impure black produces all blacks but not all of these breed true for black, for when bred together some blues appear and these prove true breeding blues.

True-breeding black* crossed blue produces all blacks, but these bred together produce progeny in the ratio of 3 blacks to one blue.

Black known to carry blue, when crossed with blue, produces on an average about equal numbers of blacks and blues, the latter breeding true.

Wild grey and self colour behave in a similar manner.

In the same way one can test the genetical make-up of a

rabbit with normal fur for the character of "woolliness". Normal fur is dominant over "woolly" fur and if any "woollies" appear, both parents carry the factor for "woolliness."

Normal length of fur is dominant to other types of fur. Any breed can be "rexed" by proper matings if a sufficient number of rabbits are produced. A true breeding normal fur rabbit mated to a rex gives all normal furred progeny in the F1, but when these are mated together they give a ratio of 4 normal to one rex, instead of the more familiar 3:1 ratio.

Fat

The normal fat in rabbits is white but some have yellow fat and this reduces the value of the carcass. Yellow fat occurs mainly in albino rabbits. White fat is dominant to yellow and the only time when complete litters of yellow-fatted rabbits are likely to occur is when both parents have yellow fat. The simplest practical way of eliminating the yellow-fatted carcass is to kill bucks which throw any yellow-fatted young and subsequently to exercise continual selection. Yellow fat can only appear when colouring matter is present in the food, e.g., greenstuff and yellow maize. If the yellow fat is present it is possible to produce carcasses with white fat by substituting hay, mangolds and potatoes for fattening rabbits.

Body Size

Body size is inherited, and is a rather complicated matter. It is perhaps sufficient to say here that when giant rabbits are crossed with small varieties the progeny are of intermediate size, and if these are bred together most of their progeny are also intermediate in size, plus a few smalls and a few heavies.

* The foregoing does not necessarily apply to certain blacks such as occur from steel Dutch breeding.

RABBIT KEEPING.

C F Snow

BREEDING

BREEDING stock must always be strong, lively healthy rabbits and they must be free from moult. Breeding from stock obviously not at their best, or suffering from moult, is asking for trouble among the youngsters when they are born.

Rabbits must be mature before being bred from. The use of immature breeding stock means that the litters will lack stamina, and the parents will never make up for the check in growth caused by early breeding. Rabbits mature at different ages, according to the size of the variety. Small breeds, like the Polish and Dutch, are often fully mature at six months; medium breeds such as the Chinchilla, Sables and Rex, mature at from seven to eight months, and big breeds, like the Flemish and Chinchilla Giganta, should not be bred from until they are eight to nine months old.

Although it is sometimes stated that a doe is almost always ready to accept service, it is a fact that wild rabbits rarely breed through the winter months. The ovaries seem to remain quiescent through the cold weather. This does not apply in equal measure to the tame rabbit, which can be bred from at any period of the year, provided the hutches are warm, and the animals are well fed, but not too fat. Over-fatness is one of the most frequent causes of unfertile matings because the breeding organs are amongst the first to lay on fat, and this makes the doe sterile.

The doe which is willing to be mated usually shows a purplish and rather congested sexual organ. If the sexual organ looks dry and pale, it is a sign that the doe is not "in season." Other signs that a doe is anxious to be mated are restlessness, and stamping on the hutch floor. Some does will carry hay about as though to make a nest, or even begin to pluck themselves. If a doe is mated immediately she begins to show such signs, the mating is nearly always successful. On the other hand, a doe may show none of the obvious signs of wishing to mate, and yet mate readily and with successful results.

If possible choose a warm, sunny day for mating stock. Does will often refuse to mate on a bad day, especially if there is a cold wind. For the actual mating, always put the doe into the buck's hutch. Never reverse the procedure, because does are often resentful of a buck being put in with them, and may turn on him and savage him before he can be removed. If the does is willing to be mated, she will at once lie in the correct position, and raise her hindquarters. If she is not anxious to mate, she may refuse to raise her hindquarters and tail at first, but after a few minutes of courtship by the buck, she will assume the correct attitude. If she refuses in spite of the buck's efforts, and remains crouched in the corner of the hutch, it is best to remove her, and try again on another occasion.

On completion of the mating, the buck will usually fall over on its side, the fall being backwards or sideways. The doe should be removed as soon as the mating has been accomplished. There is no point in allowing a second or third mating, and this is probably what will happen if the two rabbits are left together in the hutch for any length of time. The male produces millions of spermatoza at a single ejaculation, while only a limited number of ova are sufficiently ripe to be discharged from the ovaries of the doe. It is the number of ova which control the size of the litter, although litter size is indirectly controlled by the male as this is definitely an inherited character.

Does which are difficult to mate can sometimes be brought into season by feeding a few sprays of flowering groundsel, or some sprays of parsley. If this fails, soak a few maple peas (these are found in pigeon corn) until they begin to sprout, and add three or four to the ordinary food for a few days.

Another way of bringing a difficult doe into season is to put her into a hutch previously occupied by a buck, and leave her there for a day or two. Do not clean the hutch out first.

Mated does need no special treatment, but they do not want to be handled too much, and towards the end of their pregnancy sudden unusual noises are bad for them.

It is not always easy to tell if a doe is in-kindle except towards the end of the time. The gestation period is approximately thirty-one days, but thirty days or thirty-two days is no uncommon time. The shortest gestation period known to have produced live youngsters is twenty-six days, though there are several instances of does which kindled many days later than the usual period.

Pregnant does should be well fed. They do not need to be fattened, but should have the usual amount of good mash containing the best of the house-scraps available. In addition they should have plenty of succulent food to help in forming a good milk supply. Carrots are the best of the root crops for pregnant and nursing does, though others are satisfactory. Pregnant does should have the pick of the greenfood available and should always be given water, or milk and water, to drink.

It is best to use a nest box, particularly if the litter is born early in the year when the weather is cold. This should be put in about a week before the doe is due to kindle, so that she has plenty of time to get used to it. She should also be given a good supply of soft hay for nest-making purposes. Rough hay will do, but straw is unsuitable, it is too stiff and hard to be suitable for nest-making. The nest box can be filled with hay when it is put in the hutch.

Any wooden box, about eighteen inches long by ten inches wide and five or six inches high will do. A big breed will naturally need a rather bigger box. One side can be lowered a little to allow the doe to get in and out with ease. If a few small holes are bored in the bottom of the box, and two pieces of wood nailed along the bottom to raise it from the ground, this will keep the bottom of the box from becoming damp.

Some does will make their nest behind the nesting box, or even in the opposite corner of the hutch. If this happens a few days before the doe is due to kindle, the nest can be picked up and put in the box, having first put the box in the place where the nest was made. If the doe still refuses to make her nest in the box, it is best to leave her alone.

The doe will line her nest with fur which she plucks from her chest and flanks. It is quite a good plan to keep a "fluff box" into which any surplus nest-lining can be placed, so that if a doe does not make a nest, some of this spare fur can be used to keep the youngsters warm until the doe attends to this matter herself.

It is quite usual for a doe to go off her feed a day or so before she is due to kindle, and for her droppings to be rather soft. This is quite normal and nothing to be worried about. Does are always very thirsty at kindling time and a doe due to kindle must have plenty of drinking water. Lack of drinking water at kindling time is a frequent cause of does eating their young. Does need no help when they are kindling and are far better if left alone. It is always possible to tell when a doe has kindled, both by her appearance and by looking at the nest where the movements of the young rabbits can usually be discerned.

It is advisable to look at the babies when they are a day old, but the doe should be removed from the hutch while this is being done, otherwise she may be upset. If she is put into a spare hutch and given a tit-bit of greenfood or roots to keep her occupied, she need never know the nest has been disturbed.

Rub the hands well in the hutch litter before touching the nest, so that no strange smells upset the doe when she returns. Open the nest carefully at the top, and examine the youngsters as quickly as possible. Baby rabbits are born blind, and with very little fur, so that they quickly become chilled. If there is a dead youngster, or one which is obviously puny, remove it at once.

A doe should not be left with more than six or seven youngsters if she is to feed them well, and rear healthy, robust youngsters. If two does are mated at the same time, it is often possible to foster youngsters from an over-large litter to a litter which is smaller. If no foster-mother is available, it is usually best to keep more does than bucks, unless, of course, it is possible to tell at this early stage which are the best looking of any particular breed. At about a day old, baby does can be distinguished from young bucks, because of the teats which show as tiny white marks on each side of the belly. Once the fur begins to grow, these marks are no longer discernable.

If you have two does kindling at the same time, and want to foster the young of one doe, it should be done in much the same way as one examines a nest of youngsters. Fostering can be done up to four days after birth, and in some cases has been done successfully even later, but the safest time is when the young rabbits are about twenty-four hours old. There should not be more than two days' difference in age of the litters that are to be mixed, otherwise the doe may detect the difference.

Both does should be taken from their hutches, and housed separately with some food to keep them busy. Take the surplus youngsters from the large litter and put them in a box filled with warm material or cotton wool. Keep them covered otherwise they will get chilled. Cover up the youngsters left behind, then put those in the box into their new home, covering them well.

It is advisable to leave the foster-mother out of the hutch for a short while, until the newly-introduced babies thoroughly absorb the smell of the nest.

Record any changes of litters on the hutch cards. By having two or more does to kindle at the same time, it often means that youngsters from large litters which would have to be destroyed, can be saved and reared and wastage prevented.

Once the litter has been examined and everything necessary has been done, it is not advisable to disturb the nest again. The doe will do all that is necessary in feeding and caring for the youngsters, though she will not often be seen feeding them.

If the weather turns colder while the youngsters are in the nest, it is a good plan to put a little extra hay into the nesting compartment in case the doe wants to add to the nest.

If there is no suitable nesting material, the doe may pile wet litter, or even food and water dishes on the nest in an effort to keep the young rabbits warm.

The young rabbits will open their eyes when they are about ten days old, but they will not leave the nest until they are about three weeks old. If they come out much earlier than this, it means that they are not getting sufficient milk from the doe, and have been driven out because they are hungry.

To prevent this, nursing does should be given ample supplies of succulent food, particularly greenfoods. When the young rabbits leave the nest and begin to nibble at the food in the hutch, it is almost always greenfood they eat first. If the doe has had plenty of greens, the young rabbits will not be upset by their first meal of solid greenstuff.

It is as well to see that the young rabbits are healthy and free from any defects or deformities when they leave the nest. Sometimes their eyes fail to open completely by the time they leave the nest, due to colds or eye infection, and they need to be gently bathed with tepid water, and a little golden eye ointment applied.

Once the young rabbits leave the nest, particular care must be taken to see that the food is fresh, of good quality, and that it does not become soiled. The quantities of food given will now need to be increased almost daily, and any sudden changes of diet must be avoided, otherwise the youngsters may get digestive troubles, and suffer a check in growth.

The young rabbits should remain with the doe for about six weeks during summer weather, and seven weeks during winter. At the end of this time, it is best to remove the doe to a new hutch, and to leave the youngsters where they are for a week or ten days. This allows them to become used to doing without their mother before they have to be parted from each other and get used to the new hutch. All these things at once prove a big strain on youngsters, and weaning can prove a critical time unless handled with care.

At the end of a week or ten days, the young rabbits can be put into other hutches in batches according to the size of the hutch, or can be put into indoor colonies, or Morant hutches outside if the weather is good.

The doe will need a rest of about a fortnight after the youngsters are weaned. If she does not seem in good condition at the end of this time, she should be given longer. Does vary a good deal in condition after suckling a litter, but should never be re-mated until they are perfectly fit.

If a doe loses her litter at birth, she can be re-mated almost at once. It is rearing a litter, not actually producing them, which is the greater strain on the doe.

Pseudo-pregnancy.—This state of pseudo-pregnancy, or false pregnancy, occurs fairly frequently in rabbits. The doe mates quite readily, and appears to be pregnant, but in fact will produce no litter, because conception has not taken place. The doe may commence to make her nest, and her milk glands will begin to swell at about the eighteenth day after mating. This early nest-making, when the doe carries hay about in her mouth, and begins to pluck herself less than three weeks after mating, is almost always a sign of pseudo-pregnancy. When this occurs the doe should be mated again immediately. She will accept service readily at such a time, and the matings are frequently successful. Even if she should prove to be truly pregnant, such a mating will do no harm, and if it was, as is most probably the case, a false pregnancy, valuable time will have been saved.

Scattered litters.—If a doe scatters her youngsters over the hutch floor, and seems to take no interest in them, it is sometimes possible to save them if they can be found before they get too chilled. They should be put in a box lined with some warm material and put in a warm room fairly near the fire. They can do without food for twenty-four hours, but they cannot do without warmth. The cause of a doe neglecting her young is often because her milk supply has not begun to function. If the milk comes within a few hours of the birth of the litter, as it frequently does, the youngsters can be returned to the nest and the doe will soon begin to look after them. Maiden does are more likely to scatter their youngsters than experienced does.

If a doe dies, it is possible to rear orphaned babies from quite an early age by hand-feeding if one has the time to do it. Make a mixture of two parts cow's milk and one part water, and warm until the chill is taken off. Feed the youngsters by means of a fountain pen filler. A piece of bicycle valve tubing on the end of the filler makes it easier for the young rabbits to suck and does away with the danger of biting off the end of the filler. The young rabbit will suck in the normal way, and will take one or two fillers' full to begin with, but the amount should be increased almost daily. At first the youngsters should be fed every three hours during the day. They must be kept indoors in a warm room, in a box or similar receptacle, and should be lightly but warmly covered with soft woollen material or soft hay. When the youngsters are about three weeks old, they can have bread and milk from a saucer and some soft, good quality hay.

Three litters a year, and possibly four if the doe is in excellent condition, are all that should be expected from a doe. To try to get more litters than this means that the doe will have no time to recover from one litter before being mated again, and the result will be weak litters that will not thrive and will be very prone to disease.

There is a belief among some rabbit-breeders that if a doe has been mated to a buck of another breed, and produces a cross-bred litter, she will be spoilt for true breeding purposes. This is not so. Each mating affects only the litter born as a direct result of it and has no effect whatever on subsequent matings and litters. If a pure-bred doe has been mated to a buck of another breed, and produced a cross-bred litter, she will still breed true when she is mated to a buck of her own breed.

SMALLHOLDINGS - TWO ACRES AND FREEDOM.

J. O. Baker

A Few Factors Favouring Rabbit Breeding

Wild rabbits are already somewhat of a rarity, and the advantages to farmers in keeping rabbits down in numbers being so very obvious, it can be taken for granted that little, if any increase in supplies from natural sources will occur. Conversely, the demand for rabbits, both for their flesh and fur will not decrease, and can be anticipated to increase for a number of years to come. This state of affairs has created a great opening for the breeder of domestic rabbits.

A well-reared, domestic rabbit of healthy stock is far better eating than its wild brother or sister. Then, again, the value of the pelts to be obtained from rabbits must not be overlooked. These, according to breed and quality can be made to produce varying incomes. Good fur varieties being bred, and being properly managed, they provide pelts which return a handsome profit.

As a source of supply for fur, the domestic rabbit has only of recent years come into its own. That it has come, and come to stay, is certain. The real fur types provide an excellent quality fur which can compete with any other furs on the market, and entirely on their own merits. It must not be overlooked that the world supply of what might be termed natural furs, has been, and will keep on steadily decreasing. Striking example of this can be found in Canada and the U.S.A. Many wild animals, native to these countries are now never, or only rarely seen, excepting in the national reserves where they lead protected lives.

An additional source of income which must not be overlooked, is that of rearing stock for sale. Both stud bucks and breeding does. Good breeding stock realises attractive prices, and world markets in the post-war years will create an extensive demand. The English-bred rabbit is acknowledged to be the very best in the world, and foreign breeders rely on this country for their breeding stocks.

Varieties to Breed

It is possible to breed for meat only, or for meat and fur.

Conditions of space available and the facilities available for housing may influence a decision as to which line to follow. Rabbits bred for meat only are ready for the table at around four months old, and up to this age they can with safety, and with advantage, be run in "colonies." Rabbits bred primarily for their fur are not ready for disposal until six or more months old. As they cannot be kept in colonies after they reach 4 months, it is necessary to provide each rabbit with an individual hutch. This takes up space, and it also costs money.

It is suggested to the new comer that breeding for meat only should be tackled on a large scale first, with, possibly, a very limted number of a fur variety or varieties which can be increased as experience is gained, and as cash becomes freer to invest in the more expensive housing equipment demanded by the fur varieties as they pass from the colony stage.

Recommended Varieties and Breeds

Varieties which are in good demand, and which present no exceptional difficulties in the breeding and rearing of them are listed below:

Fur and Meat

Blackrex, Bluerex, Chinchillarex, Erminrex, Havanarex, Lilarex, Nutriarex, Sablerex, Argente, Beveren—Rex varieties.

Chinchilla, Havana, Lilac, Sable, Silver Fox—Normal Fur varieties.

Meat only

Flemish Giant crossed with—Belgian Hare, English, Dutch.

Dutch crossed with—Belgian Hare, English.

Belgian Hare crossed with English.

Whilst on this question of varieties and breeds, it is a matter of the utmost importance that breeding does and stud bucks are the very best obtainable.

One *may* be lucky and pick up an animal or two from which to breed, for a shilling or two, but it is highly probable that such a method of buying stock would end in disaster. It is a necessity to buy sound, healthy stock, with a parentage on both sides which have consistently produced first-class results. Such stock can only be purchased from first-class breeders, and from breeders of high repute. A really first-class firm of long standing and high repute, and which can be recommended to those wishing to make enquiries, is that of Messrs. Goodchild Brothers, Black Corner, Crawley, Sussex. This firm are advisers to the British Rabbit Council, and Mr. Claude Goodchild, one of the directors of the firm is regarded as the leading expert in this country on rabbits and everything connected with them.

Breeding and Rearing

The following brief particulars will provide a guide as to breeding and rearing operations:

Doe can be mated when 7 to 8 months old.

Period of gestation, 31 days.

Litter in summer, 6 or 7.

Litter in winter, 5 or 6.

Wean in summer at 5 or 6 weeks.

Wean in winter at 7 or 8 weeks.

Doe will give 4 litters per year for 4 years.

Breeding Rabbits

One buck to 6 or 8 does is satisfactory.

Give the doe a rest of six weeks between birth and next mating.

Inbreeding can be (and is) practised with healthy stock, but unless expert advice can be obtained, stick to normal methods, i.e., breed from parents of distinct stocks and without close relationship to one another.

Do not mate rabbits when in moult.

With fur varieties, the best skins are obtained during the period October-April. Generally speaking fur rabbits are rarely in first-class condition excepting for a very short period in other than that just mentioned, i.e., October to April. Skins which are taken when the animals are in moult are practically valueless for the better uses to which they can be employed.

Intimate details on buying stock, breeding, feeding, killing, skinning, and the like can be obtained from various admirable textbooks and pamphlets, which are mentioned in Appendix "A." Expert advice can be obtained from the breeder from whom stock is purchased, and also from various associations.

SOME INTENSIVE SIDELINES OF FARMING.

George Morland

Breeding.

Although rabbits will breed throughout the year it is advisable for the novice to begin operations in the spring, as this is the natural season. One of the secrets of successful breeding is to have both does and buck in first-class condition; a condition that can only be secured by judicious feeding and management. On no account should the buck be neglected, neither should he be allowed to become too fat.

When mating a doe of eight to ten months a buck of eighteen months should be used if possible, but one the same age as the doe will serve the purpose, provided it be well grown and full of vitality. The doe is known to be in season by her restlessness, stamping and thumping her hind feet on the ground, scratching at the corners of the hutch, carrying about a mouthful of hay, a show of temper and plucking the fur from her chest. She should be placed at once into the buck's cage: never vice versa. Directly mating is over the doe should be removed.

A doe that is backward in coming into season may be quickened by being placed into a cage recently occupied by a buck, or she may be given a feed of maple peas that have been soaked for a couple of days in water and then allowed to sprout. A handful on three or four mornings in succession will prove sufficient.

The period of gestation is about thirty days; in exceptional cases it may be extended to thirty-two days. A pregnant doe should be disturbed as little as possible and fed in the manner already described. A little more than a week before she is due she will be seen to pluck the fur from her flanks, belly and chest and line her nest with it, but plenty of hay should be available for the same purpose. During this time the doe will be working hard at her nest and will require plenty of drinking water; this should be renewed three times daily, and if a little milk can be added thereto—goat's milk is extremely beneficial at this time—it will prove advantageous.

Clean the hutch out thoroughly three days or so before the litter is expected, as this cannot be done again for some weeks. Before doing this remove the doe; grasp her firmly by the ears with one hand, placing the other under the haunches so as to take the weight of the body.

Leave the doe alone the day the litter is expected, as it is better not to look into the nest for a day or two. If there be any dead the doe will probably bring them into the run compartment. When making an examination of the litter, get the doe into the run by offering her some special titbit which will keep her busy for a few minutes. On the other hand, remove the doe from the hutch, after rubbing the hands with a little of the floor litter to take away the human smell, and give her something to eat. Inspect the nest, remove any dead, or take away some if the litter be too large for the doe to rear, and remake the nest so that it is as nearly as possible like as the one you found, give the doe another dainty bite and replace her. Any surplus youngsters may be given to a foster mother, but before doing this hold up one of the foster mother's babies and allow a little urine to drip on to those which are to be introduced to her nest. At the end of three weeks—sometimes as early as two weeks—the youngsters will leave the nest and be seen running about the run compartment. If the doe show any desire to be remated while still with her litter this should be done; otherwise in her restlessness she may injure her youngsters.

Although the youngsters will be seen to nibble their mother's food as soon as they come out into the run they should not be weaned until they are six weeks old in summer and seven to eight in winter. During the three weeks prior to their weaning, however, they should be given a little food of the kind they will receive later. This accustoms them to the change and also relieves the mother. Any lack of food at this time means a severe check in growth and, therefore, a decrease in profits. When weaning, remove the mother from the hutch and not the youngsters, as removal at this time may also cause a check in development.

The Growing Stock.

The table rabbits should be ready for killing when three months old; not later than sixteen weeks. The longer killing is delayed the larger the food bill and the smaller the profits. Careful, regular and abundant feeding are of the utmost importance. Young rabbits increase in weight rapidly, sometimes by as much as half a pound a week; therefore when half-grown they should be removed to a larger hutch and occasionally be given a run out on the floor of the shed. If new blood is wanted one or two of the best bucks may be reared to maturity and exchanged with another breeder, or they may be sold and fresh ones purchased. Stock youngsters should be separated according to sex when fourteen to sixteen weeks old.

Breeding Rabbits

Hints on Winter Breeding.

Rabbits will breed all the year round provided they be housed warmly, the stock is in perfect condition, the food supply is ample, the bedding warm, and large litters reduced in size. In very severe weather the hutches may be warmed by means of a hot-water bottle. Maiden does should not be mated after the end of October, as they generally prove indifferent mothers; only does that have already been bred from should be used for winter breeding.

THE PRACTICAL RABBIT KEEPER.

Cuniculus

BREEDING.

THE rabbit is wonderfully prolific. We have known rabbits to throw an extraordinary number of young ones in one year, and there can be no question that the power of reproduction is possessed to a marvellous extent. Twenty does and two bucks, if allowed to breed to their hearts' content, and their progeny as well, would crowd the largest rabbitry in England in the course of twelve months. From experiments we have made, we have arrived at the conclusion that the age of puberty in rabbits is about sixteen weeks in the female and eighteen in the male. The youngest period at which we managed a successful union was when the joint age did not quite amount to thirty-five weeks. Of course this was only as an experiment, and the animals experimented on were merely the chance offspring of a couple of nurse does. Six months is an age at which rabbits will breed readily, and many hundreds of litters have been brought into the world by does certainly not exceeding that age.

But no doe should visit the buck until she is eight months old, and no buck should come into regular stud use until he is at least 50 per cent, older—until he has been in existence twelve months. He may visit a doe occasionally before that period, but never until he is nine months old. If he is kept by himself until he is a year old his distinctive qualities will have developed tremendously in the meantime, especially if he is kept within sight, smell, or hearing of does. After a year he may be used freely, and will retain his powers for about four years if well kept and moderately employed.

We have said that the doe should not visit the buck until she is eight months old. Ten months is better, and under certain circumstances a year is none too long to wait. If the feeding department is well managed the extra cost will not be so large as might at first be apparent, and the power to propagate will be retained a much longer period. Does kindled any time between February and June may visit the buck early the following season, and these are generally found the most profitable rabbits, owing partly to the mildness of the weather, and partly to the abundance of green stuff to be had while they were being reared. A doe will generally be of use as a breeder for fully three years. After that time her young will not be found, as a rule, to be very fine animals. As each year she will probably have four litters, there should not be much grumbling if she is only in use for three years, seeing that in that time she will probably have given birth to between seventy and one hundred young ones. On the other hand, it is bad policy to over-use a buck, seeing that in his term of office as lord of the harem he will probably be the father of five or six hundred.

As to the number of litters that each doe will have in the year, this should be limited to five at the very outside. We prefer four, and believe it to be the most profitable number. The best months then will be February (middle), May (early), July (late), October (early). There is thus four months rest left for winter, a period during which breeding is seldom found to answer very well, for reasons which will be at once understood. Several breeders go in for five litters a year, some even for six; but were the numbers of young to be fairly compared we believe the balance would be on the side of four litters, on account of the large number that die when the doe is too heavily taxed; quite independent of the increased strength and value of the progeny.

The times we have named for breeding will be regulated to a great extent by the doe, who will come in season at uncertain periods. Some does are in season immediately after kindling; others, on the other hand, will not receive the buck for three or four months after. It is difficult to say which of these two is most objectionable. If the doe comes in too soon she may suffer from a complaint we shall refer to in our chapter on diseases, and if she delays too long much valuable time is, of course, wasted. Well-fed and well-cared-for does seldom give any trouble, although there are exceptions to the rule. We once had a doe that only consented to be paired three times in two years, and which only threw seven young ones during that period. We bore with her patiently for two years, and then fattened her for the pot, finding that she would not pay for her keep. This, as we have said, is a very exceptional case. A doe when about six months old generally comes in season if the weather is warm, and at about eight months in very cold districts. The symptoms are an enlargement, and generally slight inflammatory look about the external organs, a rather wild look, and a general restlessness of demeanour; also a liking to be stroked or fondled, and frequently a habit of tearing up the bedding and carrying it about in the mouth. These are very critical periods, and the doe should be fed "low" for about a week. All heating food, such as barley or turnips, should be avoided, and the supply of corn should be limited. Cool, succulent, green food should be the principal article of diet, and the rabbit will usually be right again by the end of a few days. She will probably come in season again after about a month, and a similar course of treatment must be followed.

When you have decided to allow the doe to breed, select a buck, and place the doe in his hutch, which should have no sleeping compartment, or if it has, the door should be let down. Do not attempt to force the doe, and do not let the buck worry her: if she declines his advances and shows fight, remove her till the following day, when all will probably be right.

There is one point of great importance with reference to this matter, and that is that the buck should only fall once. Frequently he is allowed to repeat the act three or four times, under the delusive idea that the number of young ones will be increased. A more delusive idea never obtained belief. We have kept careful records of this matter, especially for the benefit of young beginners, and give the following nine results with the same buck and two does:—

Number of Falls.	Length of Gestation.	Number of Young.	Number Reared to Two Months.
3	31 days.	3	1
3	30 days.	2	—
4	31 days.	2	1
5	No result.	—	—
5	,,	—	—
2	30 days.	4	3
2	29 days.	3	3
1	30 days.	8 (2 killed)	6
1	31 days.	6	6

These results really show about the average; and, quite apart from the harm done to the buck, it will be seen that one connection is much the most successful. Into the physiological reasons for this it is not necessary to go in this treatise, which is purely a practical one.

Breeding Rabbits

A week or ten days after the union the two should be placed together for a second or two, and if the doe is fractious and pugilistic there need be little doubt but that she is in kindle. Avoid handling a doe in this state. There is nothing more foolish than to pull about a rabbit at any time, especially during the most critical periods of her existence, and we have no patience with people who feel a rabbit every day, "to see if she is all right." Nothing on earth is more likely to make a rabbit go wrong. On one occasion a man professing to know all about rabbits felt a doe of ours during the first few weeks of our experience. We had rather rashly bought her while heavy with young, and asked the individual referred to to feel her, and ascertain if all was right. He did so, and told us we had been swindled. The next morning the poor animal had given birth to eight young ones, and had died in the delivery, having evidently received some internal injury. It is all very well for a fancier of experience to take his own course to ascertain if "all is right," but a young fancier should be content to wait for signs. At the end of about the twenty-fifth or twenty-sixth day the doe will generally show some signs of activity in the matter of preparing for her expected family. Sometimes this is delayed a day or two, and sometimes it is attended to a week before the delivery. A good supply of soft hay and straw should be given out now, and with this the expectant mother will make a warm nest, lining it with fur dragged from her breast and back. Instinct teaches the rabbit to attend to this matter, so as to make the little ones quite warm. As they disfigure themselves in this way sadly at times, it would be well if they could be persuaded into using a little sheep's wool or wadding, and we have tried this several times, but seldom found that the stuff was made much use of. Clean the hutch out very carefully, and supply the doe with a little cold water daily. This precaution is most necessary, as it is only a feeling of intense thirst or of extreme fear that can make a doe eat her young.

The period of gestation in rabbits is about thirty days—sometimes a day more or less, but seldom varying more than three days. Taking one year of our rabbit book, we find that fifty-two litters were born. The aggregate length of the going with young was 1,582 days, giving an average of 30.4 days each. Again, it is noticeable that some does go a day or two longer than others. Taking the year referred to, one doe carried four litters—a total period of 113 (29, 28, 28, 28) days; two others, 119 each; three others, 120 each; five others, 121 each; another five litters, 161 days; and the last three litters, 100 days (32, 34, 34). The first and the last are both exceptional, the last especially so. We may state, however, that we adopted a plan which made a mistake a moral impossibility. The diary was entered up daily, and in addition to this a small slate (about three inches square) was tacked to each hutch. On this was the number of the rabbit contained in the hutch, and in case of service of doe, the buck was given, with the date of service, and of the expiration of twenty-six days, at which extra precautions were needed. When completed, the tablet read thus:—

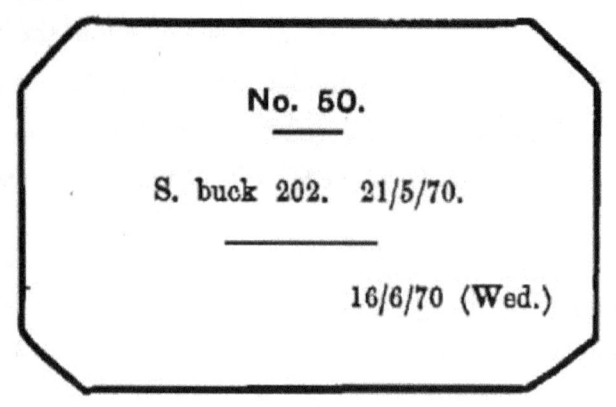

```
No. 50.

S. buck 202.   21/5/70.

                16/6/70 (Wed.)
```

Then as soon as the doe kindled the two bottom items would be rubbed out, and simply the date of birth substituted. The trouble was a mere bagatelle, the cost not a shilling for a large rabbitry, and the convenience very great. In fact, a very great saving was doubtless effected, as mistakes were avoided.

There are several precautions to be taken prior to the rabbit's accouchement. The doe loves quiet, and is very jealous of any interference. As this is the case, and also for a little extra warmth, a sack may be nailed over about half the framework of the hutch, next against the sleeping compartment, so that the hole leading to the place is quite concealed from outside view. Often this sack is put the whole length, but this tends too much to check the flow of air, and is not on that account to be recommended. Any sudden noise, such as the violent banging of a door, loud shouting, or stamping, may bring on premature delivery, and of course death to the whole family. Sometimes a violent thunderstorm even will have this effect. It is impossible to guard against this, but it is possible to keep the rabbit as quiet as possible, and to ask visitors not to make any great noise. Dogs and cats should never be allowed to enter a rabbitry. Sometimes these animals are so well trained that there would be little or no danger of their hurting anything, but they are very apt to frighten pregnant does, and thus to do considerable harm. Even such pigmies as mice will sometimes frighten a doe out of her wits.

When first we commenced rabbit-keeping we accused a doe of eating three litters of young ones in succession, and eventually sold her to a medical student who made a hobby of rabbits as well as of hospitals. He experienced the same occurrence, and could get no litter from the doe at all. We clubbed our heads together, compared notes, and mutually wondered how it was that the doe cleared off her young on each occasion so neatly that not a spot of blood or a single sign could be found. We determined to watch the next time, and did so, finding that the doe really did not kindle at all. Each time she received the buck apparently with great readiness, refused him when tried after a week or ten days, her teats swelled (an almost certain sign of pregnancy), there was an evident secretion of milk, and as the month was expiring she made her nest with scrupulous regularity. It was evidently a case of sham pregnancy, which we believe is confined entirely to the rabbit race. Since then we have come across the following paragraph in "The Rabbit Book,"* which, being the only published information on the subject we have been able to find, we reproduce:—

"Rabbits that have not become pregnant frequently offer the physiological phenomena of a real gestation. They prepare their nest, their teats swell, and the secretion of milk takes place. They have been known to adopt, when the opportunity offered, strange young. Harvey, the immortal discoverer of the circulation of the blood, says that he has seen and observed this physiological phenomenon in rabbits." Since that time we have occasionally come across such a case, but they have not been very frequent. The phenomenon is most remarkable, and whenever it occurs causes great trouble to young fanciers. Many a good doe has been accused of eating her young ones, and destroyed in consequence of it. It is by no means incurable, good sound food and occasional exercise being the best remedies we can suggest.

The usual sign of pregnancy, increased size of the belly, is not to be relied on in rabbits; because, in the first place, the young are carried so high that they sometimes do not affect the mother's appearance at all; and in the second place, because many of the ailments to which rabbit kind are subject show themselves in a swelling in this part of the body. Hence it is not so easy to detect this singular occurrence as it otherwise would be.

The food of the doe at this time should be most carefully selected. The green stuff should consist chiefly of endive, lettuce, milk-thistles, and, in fact, all kinds of the most herbaceous nature, in order to promote the secretion of milk. The dry food should also be freely given, and should consist of a feed of oats daily and a liberal supply of meal mashes. Just at the time of kindling barley-meal mash should be given, as it is of a warm, comforting taste. It is best given warm, and a little milk added is not thrown away. The usual basis of feeding should not be followed just now, as the rabbit will have shortly a great strain on its constitution.

As a rule, it is unnecessary, as well as unwise, to interfere with the nest as prepared by the mother, who knows a good deal more as to the requirements of her expected family than a fancier himself can possibly do. There are, however, occasions when it becomes necessary to alter the position of the nest; such as, for instance, when a doe selects a damp corner of the hutch, or when a violent storm—in outside hutches—has drenched it. When this has to be done, let it be done quickly, and be careful to move the whole bodily, bearing in mind that if the doe finds her work disturbed she may very likely desert

Breeding Rabbits

it entirely.

The act of parturition is not accompanied by any great suffering, nor, if the rabbit is in good health, inconvenience. The young are born singly, sometimes very rapidly one after the other, at others the interval being very long. We never knew a rabbit die from exhaustion on kindling, although we have seen some reduced very low indeed. Good and careful feeding will almost invariably restore the animal to its old strength. While the rabbit is being delivered it should be kept perfectly quiet, and not disturbed in any way; but it is important that the water referred to should be given, or great thirst may be the result. After the birth it *may* be necessary to look at the young rabbits, and even to touch them, but neither should be done except in cases of actual necessity. During the second day the sliding door into the living room should be shut, and a peep taken into the nest. Count the little creatures without handling, and remove with great care any dead ones. If there are over six young ones, two or three must be removed if special excellence is aimed at, or the whole will be but of moderate size and strength. If any are much smaller than the rest—there is often a "darling," as in the case of pigs—it is as well to destroy them, provided there will be a good number left, as the small ones are terrible blood-suckers, and take a lot of food which ought by rights to be consumed by larger and more valuable members of the community.

It is difficult to say how many young ones a doe may bring up, consistent with all being good show animals, but we should say four were the very utmost in the case of Lops, and perhaps five in the foreign breeds. As on several occasions the does will probably throw seven or eight, it will be a pity to lose the surplus ones, and for this purpose nurse-does should be kept, and one of these should kindle simultaneously with each doe of value. Dutch does make splendid nurses, and so do several of the common varieties. We have found a cross between the two act very well indeed. In a large rabbitry there will be sure to be two or three badly-marked Dutch does. These may be used for breeding, and occasionally a very nice young one will be the result. As a rule, however, the young will not be up to much, and these may be destroyed and their place taken by valuable animals. The process of changing must be carried on very carefully, or the doe will discover the fraud, and either destroy or starve the new-comers. It is best to adopt a time when the doe is feeding. Then slip the partition down, and change rapidly, covering up the little ones just as they were found. There need be no fear of suffocation, although the manner in which the mother frequently heaps up the bedding on the top of her progeny may cause some little alarm to the beginner. Keep the door closed for over an hour, during which time the young rabbits will have got mixed, and the difference in smell will not be so apparent. The mother, during this period, will get somewhat heavily overloaded with milk, which she will be anxious to get rid of, and in her hurry to do so the change will be overlooked. But there is always considerable danger in these matters, and in order somewhat to obviate this it is better, if possible, to leave at least one of the old litter to the new-comers for a few days.

The nurse-doe should not be too heavily taxed. As a race they may be called an uncomplaining, hard-working lot, but this is no reason why they should be made prematurely old, or why the young ones should not have a fair chance of obtaining nutriment. Sometimes the young are left with the doe until they attain a great age, perhaps even three or four months, but this is not to be recommended if carried to excess. Sometimes, again, after a nurse-doe has reared a litter for six weeks, these are taken away, and a fresh litter of newly-born young ones substituted. In this case the change has to be made very briskly, and the rabbit must be fed on very succulent food for a few days before the removal. These long-suffering animals will often take to a second and even third litter, and bring them up with as much care as if they were their own; but in the event of the poor animal eating her foster children, cannibal as is the act, she cannot be very much blamed, certainly not one half so much as the fancier who is guilty of such an act of cruelty. The act is an extremely cruel one, and we should only recommend it in case of a mother dying.

Breeding Rabbits

The young rabbits are born without any hair, and quite blind. After a few days a silky fur comes on them, and then grows with great rapidity. When about a fortnight old the little ones begin to crawl out of the nest, and will come to the feeding trough soon after. Great care is needed in rearing the young rabbits, as much loss is often experienced owing to accident and other ways. Ground oats should be given them as soon as they begin to eat. Too much green-stuff should be avoided, and if what is given is placed in the rack the youngsters will not come in for too much. As soon as the young rabbits begin to come out of the nest, the whole may be removed, and a fresh supply of bedding, in the shape of dry hay and straw, substituted. This will be found to be necessary, as the young ones will have made the original nest smell decidedly unpleasant by this time. A good plan is to remove the lot bodily into a clean hutch, and then allow the old one to be ventilated and whitewashed.

A most important matter to be observed at this time is the feeding of the doe. When there is a large litter the tax upon her strength is tremendous, and she should be fed accordingly. Corn should be given *ad lib.*, and almost any amount of succulent green-stuff may be supplied. Any evil from too much of this will be guarded against if plenty of fresh sweet hay is also administered. It may safely be taken that the doe should have about twice the ordinary supply of rations at this time, and the youngsters must be watched. As fast as they commence feeding the supply must be increased accordingly. We found it well when there was a large litter, say over three, to shut the little door, and allow the mother to eat in peace, instead of letting them worry her the whole time. This should be done especially when she has her mash supplied, which should be bread and meal alternately, and it may be mixed a little more sloppy than usual. Boiled potatoes mixed with meal or ground oats we have found very palatable to the doe at this time, and is a pleasant warm food.

A difficulty often presents itself that while the hutch door is open, the little turks peep over the sides, and overbalance themselves, falling headlong on to the floor, and breaking either neck or limb. We once saw it stated that this was good for young Lops—we mean the peeping and not the falling—as it tended to cause the ears to fall properly. In this we beg to differ. To prevent the inconvenience it will be found handy to put a thin slip of wood along the bottom of each hutch in the front, in such a way as to prevent any falling out while the frame is open. The principle to be adopted is similar to that advocated for sleeping compartments in the chapter on hutches, except that the slip need not be so wide.

Avoid handling the young rabbits, and when your friends come to have a peep, confine their attentions to looking. A run on the floor occasionally is very good for them, and may be indulged in while the doe is having her special meal. The gambols of the youngsters are most amusing, and the results are strengthened limbs and improved constitutions.

When about three weeks old the rabbits pass through a moult, and another when about twice that age. At both periods, the latter especially, great care is required to prevent their succumbing to what is really a disease something akin to measles, and for that reason we never remove from the mother until about seven weeks after the birth. When the young ones are of special merit, eight weeks is the safest time, and if a week longer no harm is done. A fancier will, however, be hardly likely to allow a doe of any value to be pulled about by youngsters of more than that age. If they are, however, removed before they attain at least six weeks there is great danger of their not having sufficient strength to get over the moult.*

Do not remove all the young ones at once, as the doe will suffer a good deal in that case. Take one away at a time, and place it in a warm hutch *with some other young rabbits*. This is very important, as if the precaution is not taken, the poor little thing is likely to suffer much from cold, and nothing puts a rabbit back so much as a sudden chill. The next day another young one may be taken away, and the remainder on the following day. The doe should then have less succulent food, but should be kept well supplied with strengthening and fattening articles preparatory to next litter. A little salt should be scattered with the corn, and also a little flowers of brimstone. The result is to dry up any milk that may be left, and cleanse the stomach from any impurity. Supposing the young ones to have been removed at the end of the eighth week, the doe will have to visit the buck again almost immediately, so that it is best to save a few days if they are pretty forward. The doe should be fed high as stated, and she will probably come in season again at once.

Many of our readers will say that we are giving the young too long with the mother, and thus losing time; but it should be remembered that we are dealing only with fancy rabbits in these remarks, and that it is far better to rear a dozen good animals than twenty bad ones. The former will fetch fancy prices, and take large prizes; the latter will be practically worthless, and often not worth the trouble of rearing. Many fanciers pair the doe directly she comes in season, whether she have young ones with her or not. For fancy rabbits this practice is to be much reprehended, however much it may be allowed where numbers and weight for the pot are considered the only matters of importance. Sometimes a doe will take the buck the day after kindling, and we once knew a poor animal that kindled twice in one calendar month. This was a very remarkable occurrence; and the doe died after rearing her second litter, while only one out of the sixteen young ones attempted to be reared reached the age of discretion. This was, to our mind at least, a conclusive proof of the penny wisdom and pound foolishness of overdoing the matter in rabbit breeding.

Returning to the young rabbits themselves. These should be kept in roomy hutches well ventilated, and although large numbers may be kept together, they should not be crowded. It is sometimes found advantageous to have a door made in a row in a stack, so that the young rabbits can have a run of seven feet. They enjoy racing through the whole, and will benefit by the exercise. They will eat very freely, and should have a good supply of corn, with a judicious addition of roots and green-stuff. The latter should not be given too freely, and plenty of hay should be served out daily. The good effect of this is very marked. Until the rabbits are thirteen or fourteen weeks old, they will do no harm at all by running together, but by that time it will be necessary to see about separating the bucks from the does, and the bucks from one another. In the male the reproductive instinct is pretty fully developed at four months, and therefore they should never be left with the does after having arrived at that age. They soon begin to learn how to fight, and will clearly demonstrate that it is not only dogs whose delight it is to "bark and bite." Hence they soon have to be separated, or they do one another terrible injury. The does can be kept together, three or four in a hutch, till they have attained the age of six or seven months. As a rule, there will be no difficulty, but if any one of them exhibits signs of fighting, she must be removed at once. Both before and after the separation the food must be very judiciously selected. Oats are best crushed for the first few months, nor should whole barley be given. All green-stuff must be especially dry, and should not be given too fresh.

There are one or two small matters to be attended to in breeding, but these will be better dealt with while speaking of the different varieties. The important question of breeding for colour will come under Lops, that being the only variety, except Dutch, in which colour—we mean fancy colour—is of importance. In different foreign breeds, shade is important, and a few hints on the subject will be found in their proper place.

Just a word as to strains. A great deal is said about A's strain of Lop, B's of Himalayas, C's of Belgians, and so on. Much of this is mere balderdash. A sells a young rabbit to D, who rears it, and breeds from it. Not having a name himself, he advertises the offspring as A's strain. If the rabbits are well bred, patience and a certain amount of ability will make them pretty much perfection despite anything that may be adduced to the contrary. For our own part, we always found it best to constantly introduce new blood. This we did to avoid degeneration, owing to blood relations breeding together.

In many rabbitries, a buck does service for half a dozen does. He may do so easily, and generally for eight or nine. But he cannot keep on serving the whole rabbitry, if all the young does or even a great part of them are kept. The consequence is, that when the father cannot do all that is required, one or more of the brothers is selected, and all laws of nature are broken. The result is, that frequently—more so than the average—the unions are not fruitful. The offspring themselves do not suffer materially in the first generation, but if there is any taint in the blood, it will be sure to increase and come out. But when repeated, generation after generation, without the infusion of any new blood, the result is deplorable. Advocates of in-and-in breeding may say what they will, but the result is invariably miserable, puny rabbits that are a disgrace to their hutches, and which rarely, if ever, take prizes. In some breeds the course is purposely pursued to diminish the size and weight, but where fine rabbits are required, it is to be eschewed religiously. We once visited a rabbitry of a young friend who had started with the delusive hope, held out by a small book, that he would be able, by the investment of £5, to realise an income of a pound a week. After three years, not only had he made no profit, but he had had to borrow another £5 to keep his rabbit account square, and was also in debt to a considerable extent. He managed everything apparently very well, and kept his place very clean. But his rabbits for the most part looked sickly and wretched, and we saw no less than three youngsters positively in the act of dying. We looked at his logbook, and found that he had not bought a single rabbit since he started. Commencing with a buck and two does, he had got twelve litters in the first year, and reared a good many of the youngsters. Then he sold the mothers, and bred with the sire and his offspring, and this had been going on for two years, the buck serving twelve does. We give this as a simple instance of the evil of in-and-in breeding, and also of overtaxing the energies of a buck. While the season is on, he cannot be fed too highly, and warm food should form a staple article of diet. Then he will be able to act for six or eight does nicely, and even more, provided they do not breed too rapidly. If a sure stock-getter, he will only have to visit forty does a year, if ten are kept and only four litters each, and this is not too many, even if he occasionally has a visitor from another rabbitry. There is a good deal of freemasonry among rabbit fanciers on this point, and it is very useful. Never allow a buck to serve twice in one day, or, if possible, on two subsequent days, and bear in mind the caution as to more than one union.

A question we have been frequently asked both privately and through the *Live Stock Journal* is, What can be done to make an obstinate doe come in season? Barley and barley meal seem to serve this purpose sometimes, if given exclusively, and there are more violent measures which we do not recommend for any but old hands. One is, to give a little cayenne in the

Breeding Rabbits

food; another, to administer a miniature dose of cantharides. The worst of this latter is that it is a poison, and that a small amount will kill a rabbit. Constitutions differ also so materially, that a dose so weak as hardly to take effect on one doe, would kill another. Perhaps the best way of all is to place the refractory doe in a hutch occupied immediately before by a buck, and put the latter in a position where the doe can see him. Want of condition is a common reason for refusing, and another is a disease to which we shall allude in the chapter on diseases. Under either of these circumstances the only thing to do is to remedy the defect as quickly as possible, and then to try again: as we have said before, never force a doe. It can generally be done, but seldom results in anything except souring the animal's temper, and making her afraid to enter the buck's hutch ever after.

There is a difficulty among amateurs in distinguishing the sex of the young rabbits, and this causes some inconvenience in disposing of surplus stock. It is a singular thing that the sexual difference is scarcely discernible in rabbits except to physiologists for the first eight or nine weeks, although at three months the difficulty is solved. A little experience will teach the fancier how to distinguish the sexes pretty easily; but generally speaking a pretty fair idea can be gained from the formation of the head, and the general demeanour. The head of the buck is more solid in its appearance, and wider across the skull than that of the doe; also the young buck is generally more frisky and troublesome than his sister. As, however, the organisation of rabbits differs materially, it is difficult to rely absolutely on the first test, and there are very quiet bucks and very wild does, so that in a few instances the second distinction cannot be said to be always correct.

* "Manuals for the Many." 171, Fleet Street.

* The causes of infant mortality amongst rabbits, and the best means of avoiding the evil, are discussed in the chapter on diseases.

CPSIA information can be obtained
at www.ICGtesting.com
Printed in the USA
LVHW031002110720
660407LV00003B/780